COLLINS AURA GARDEN HAN[DBOOKS]

PROPAGATION

TONY DERRICK

COLLINS

Products mentioned in this book

Benlate* + 'Activex'	contains	benomyl
'Keriroot'	contains	NAA + captan
'Rapid'	contains	pirimicarb
'Sybol'	contains	pirimiphos-methyl

Products marked thus *'Sybol'* are trade marks of Imperial Chemical Industries plc
*Benlate** is a registered trade mark of Du Pont's
Read the label before you buy: use pesticides safely.

Editors Maggie Daykin, Susanne Mitchell
Designer Chris Walker
Picture research Moira McIlroy

First published 1988 by
William Collins Sons & Co Ltd
London · Glasgow · Sydney
Auckland · Toronto · Johannesburg

Reprinted 1989

British Library Cataloguing in Publication Data

Derrick, Tony
Plant propagation. —— (Collins Aura
garden handbooks).
1. Plant propagation
I. Title
635'.043 SB119

ISBN 0–00–412384–0

Photoset by Bookworm Typesetting
Printed and bound in Hong Kong by Dai Nippon Printing
Company

Front cover: Pelargonium cuttings by Pat Brindley
Back cover: Streptocarpus cuttings by The Harry Smith
Horticultural Photographic Collection

CONTENTS

INTRODUCTION

Propagation comprises the whole range of techniques used to make more plants from existing ones. Most of these techniques imitate nature's methods. To increase plants successfully, one needs to be familiar with the various methods and also know which is most suitable for each type of plant.

Sowing seeds of vegetables, such as lettuces, beans and marrows, is a simple form of propagation which even complete beginners are bound to succeed with.

Most people who have an interest in gardening have already propagated plants, perhaps without realizing it. Sowing a row of lettuces, a windowsill crop of mustard and cress, or a patch of annuals for the summer, or rooting some tradescantia or busy lizzie in a glass of water, are all simple forms of propagation which are almost bound to succeed. But the more advanced techniques are not difficult once the principles are understood and the rules carefully followed.

One reason for propagating plants is to save money. Bedding and other plants needed in quantity for a bold display are expensive to buy: it is far cheaper to sow a few packets of seeds or root some cuttings.

Propagating is also great fun. There is great satisfaction in creating new plants from seeds, or pieces of shoot, leaf or root. It is fascinating that a tiny, dry seed has potential for such growth and holds the plant's blueprint in every detail, and equally intriguing that many pieces of shoot, leaf, or even root can be coaxed to form complete new plants.

Rooting cuttings and other techniques also make it possible to save or rejuvenate old or damaged plants difficult to replace in any other way.

There is a vital distinction, however, between raising plants from seed – sexual reproduction – and doing so from parts of the parent plant – a stem, root or leaf cutting, a sucker or division: asexual or vegetative reproduction. A plant raised vegetatively is identical to its parent as there has been no intervening sexual stage to bring about any recombination of the genetic material. All plants vegetatively produced from one original parent form what is called a 'clone'.

Seeds, on the other hand, draw the genetic material that determines the characteristics of the plant – height, flower colour, scent, or lack of it, etc. – from both male and female cells, the ovules and pollen grains, and as a result of recombination of the genes at fertilization can differ in appearance from the parent or parents. Natural species show little variation, in fact, as their parents are for all practical purposes identical. But man-made hybrids have such a complex parentage that their seeds would be unlikely to give similar looking plants. Professional seedsmen ensure that their seed strains come true to type by carefully controlled pollination and rogueing (weeding out the unwanted plants). So, in order to produce exact copies of a plant it is necessary to use one of the vegetative methods of propagation.

Most parts of plants can be used for propagation. The *Abutilon megapotamicum* illustrated can be increased from seed or semi-ripe cuttings taken in summer.

SEEDS

Seeds provide the cheapest and generally the easiest way of creating new plants. This is nature's usual method too. Seeds of hardy plants are generally raised outdoors in the open ground, but tender kinds are started in pots or seed trays in a greenhouse or heated frame or on a warm windowsill. Before looking at sowing techniques, it is a good idea to understand something about seeds themselves – particularly what makes them germinate. A seed contains the germ of a new plant, with its first root and shoot and a store of food to fuel it in its early life, within a protective seed coat.

Some sweet peas have hard-coated seeds which must be chipped so that moisture can penetrate inside the seed coat.

Use a sharp knife to chip the seed coat, making the cut on the side opposite to the eye. The shoot is produced from the eye, so do not damage it.

In order to germinate, the seed must have moisture, oxygen and sufficient warmth, and, soon afterwards, light which provides the energy it needs to manufacture food. If moisture cannot penetrate the seed coat, the seed will not spring to life; nor will it do so if the soil is waterlogged or compacted, so that oxygen is in short supply. There must be enough warmth too, in order that the chemical processes within the seed are stimulated and growth started.

All these factors work together to mobilize the seed's food store, on which it must depend until it forms leaves and roots. All these needs must be met for healthy growth.

Some hard-coated seeds (e.g. dark sweet peas) will germinate better if the seed coat is chipped with a sharp knife on the side opposite the eye so that moisture can penetrate. Many tree and shrub seeds need cold treatment before they will germinate (see Stratifying, opposite page).

Collecting seeds Most seeds sown are bought from seedsmen, but you may wish to collect seeds from your own plants or from a friend's, and there are a few useful guidelines to follow. They must be viable, that is, with enough life in them to germinate, given the right conditions. Shrivelled seeds will never germinate. (Never collect from diseased plants, either.) The smaller the seed, the more rapidly it will lose its germinating powers. Timing is crucial. Collect only fully developed pods or seed heads, usually in late summer or autumn, though some are ready earlier.

Cut the seed heads carefully on a dry day before the seeds fall. Lay them on sheets of paper or card, so that no seeds are lost, in a cool, airy greenhouse or other suitable place to dry off. Any that are likely to explode, such as those of poppy, should be put in lidless boxes.

Seed cleaning When the seeds are ripe and dry, carefully separate them from the pods and other debris. Coarser material can be removed by hand, but some seed heads have to be crushed and the seeds sieved out or the chaff shaken or blown away. Packet the seeds, label them clearly and keep them in a cool dry place until sowing time – probably early the next year, though some primulas and meconopsis, for example, are best sown straight away.

Stratifying Fleshy tree and shrub seeds, such as those of holly, need special treatment before they will germinate. Spread alternate layers of sharp sand and seeds in a plastic flower pot, protect from mice and birds with wire netting, then stand the pot outdoors at the foot of a cold north wall for six months for frost to destroy the flesh and soften the seed coats. The seeds are then sifted from the sand or sown direct. Seeds such as yew and berberis should then germinate when sown, but some, holly for example, require about 18 months of this treatment.

TOP Collect only fully developed seed heads, usually in late summer or autumn. Choose a dry day.

ABOVE Berries of many trees and shrubs have to be stratified before they are sown.

Seed Sticks
This is a recent development, available in a wide variety of seeds, in which each seed is glued to a cardboard stick, thus allowing accurate and properly spaced sowing. Simply plant the Seed Sticks to the depth marked on the stick in the prepared soil.

RIGHT Marigolds are among the many seed-raised plants that have a wide range of F1 hybrid varieties.

LEFT When sowing in drills outdoors it is essential to scatter the seeds thinly to avoid wastage and overcrowded seedlings.

SEED-SOWING OUTDOORS

Most seeds of hardy plants are sown in rows in the open ground where they are to mature, but some flower and vegetable seeds (biennials and members of the cabbage family) are raised in a seedbed and set out in rows later.

Good soil preparation is vital to create the best conditions for germination and growth. Dig the plot in winter, then let it settle. Just before sowing, fork the soil through to level the surface and shatter any lumps, tread it firm, apply a dressing of general fertilizer according to the manufacturer's instructions, then rake it sufficiently to create a fine, crumbly surface.

At sowing time the soil should be warm enough for seeds to germinate and moist but not wet. Always be guided by the temperature and state of the soil, not by the date.

Sowing in rows makes weeding and thinning simpler, so use a measuring rod to space the rows accurately and a garden line to keep them straight. Draw out 6–13 mm (¼–½ in) deep drills with the corner of a draw hoe held against the taut line. If the soil is dry, water the drill but allow the water to drain through before sowing. Don't sow too deeply or the seeds might not come up, particularly if they are very fine.

Sow thinly to ensure better growth and avoid wasted seeds and overcrowded seedlings. Techniques vary: tap seeds from a corner of the packet; take seeds between thumb and forefinger and sprinkle along the row; tip seeds into your palm then tap them off into the drill with the other hand; or use a simple seed-sowing device, available from garden centres. Sow large seeds, such as peas and beans, individually with your fingers. Rake a light covering of soil over the seeds after sowing, but don't bury them too deeply. Label the rows.

F1 Hybrids
These are offspring resulting from seeds produced by crossing two selected varieties. They are uniform in appearance and have a better flowering or cropping performance than either of the parents. However, if F1 hybrids are themselves crossed, the offspring show a continual decline in vigour and uniformity, so it is not worth saving seed from them. You will need to buy fresh from a nurseryman.

Fluid sowing This method of sowing pre-germinated seeds in a protective gel is useful for seeds that are slow to germinate or when conditions are very dry. First, sow the seeds on damp tissue or kitchen towel and keep them in a warm dark place until germination occurs. As soon as the tiny roots are visible, remove the seeds and mix them with wallpaper adhesive such as Polycell or with a gel specially prepared for the job. Put the gel and seed mixture into a polythene bag, snip one corner off and squeeze the gel into the prepared drill. Cover with soil in the usual way. The gel protects the seeds and allows them to be sown more thinly.

Thinning Space-sown seedlings do not need thinning. Thin rows of seedlings such as carrots in stages – first to a single line, then gradually to the full spacing recommended.

Seedbed method Do prepare the seedbed thoroughly to ensure plenty of sturdy seedlings for transplanting. Take out 90–120cm (3–4ft) long drills 10 cm (4 in) apart and 6–13 mm (¼–½ in) deep. Sow thinly and cover with fine soil or pea shingle. Water in dry weather, shade from fierce sun with plastic netting and keep clear of weeds. Lift the mature seedlings after watering them and set out in nursery rows or where they are to grow.

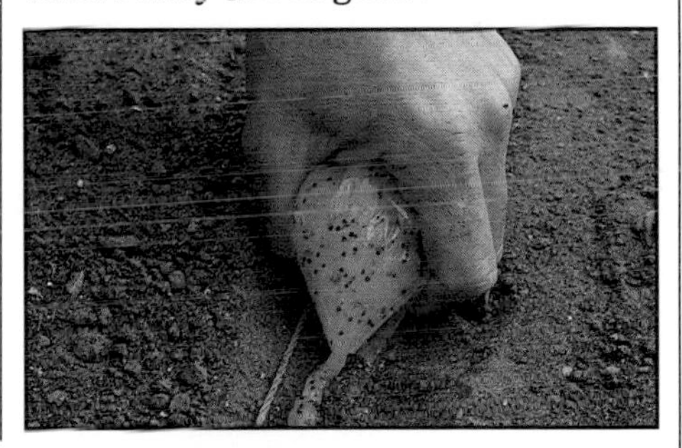

ABOVE Fluid sowing is useful for seeds that are slow to germinate.

LEFT Thin rows of seedlings in stages until the full recommended spacing is achieved. Water afterwards to settle soil.

A bed marked out in bold patches with sand, ready for sowing hardy annuals. Note the shallow drills, 15cm (6in) apart, within each patch.

Hardy annuals Sow seeds where they are to flower in March/April when soil and weather are fit. Choose an open sunny site on not too rich soil, so that the plants don't make excess foliage rather than flowers. Prepare the soil as for a seedbed, mixing in peat or similar material to hold moisture in dry weather. Rake in a scattering of general fertilizer just before sowing. Mark out boundaries of patches for each type of flower with sand, making them about 90 cm (3 ft) across and keeping plant heights and flower colours in mind, and sow seeds in shallow drills 15cm (6 in) apart within each patch. The plants will soon cover the ground.

The hardiest annuals (e.g., cornflower, calendula) can be sown in September/October for an earlier spring display if the soil is well drained. Annuals that are being grown for cut flowers are best sown in rows across the vegetable plot.

SEED-SOWING INDOORS

There is far more to sowing indoors than just giving seeds and seedlings extra warmth and an earlier start than would be possible outdoors. The aim is to control all their growing conditions to give them the best chance of success.

This involves protecting them from pests and diseases, particularly damping off disease, and using well-drained composts that provide the moisture and nutrients they need. It also means providing enough warmth to germinate the seeds and produce sturdy seedlings, sheltering them from wind, rain and frost, but with ventilation to give them air without draughts, and supplying them with water when they need it. Raising seeds in pots or trays makes it easier to keep a close watch on their progress and give them prompt attention.

Many plants are raised like this, generally between January and April – pot plants for the home and greenhouse, half-hardy bedding plants, tender fruit and vegetable plants such as tomatoes and marrows for planting outdoors later. Some hardier plants benefit from such care in their early stages so that they are stronger when eventually planted out.

Good hygiene is vital. Start with a clean greenhouse, free from pests and diseases, clean pots and seed trays, and sterilized soil- or peat-based seed or potting compost such as 'Kericompost'.

Fill 2.5–3.5cm (1–1½in) deep seed trays to within 10mm (⅜in) of the rim with compost. Firm with your fingers, especially round the edges, and level the surface with a wooden presser or the flat base of a flower pot. Stand in a tray of water until the surface is dampened, then leave to drain. Sow thinly and evenly, mixing the finest seeds with clean, dry sand to make this easier. Cover with glass to retain moisture and shade with paper until the seeds germinate, then remove it promptly.

Seeds are best germinated in an electrically heated propagator or on a bench heated with electric soil-warming cables. Half-hardy annuals, tomatoes and many others need about 15°C (60°F), the more exotic begonias and gloxinias 18–21°C (64–70°F). Excessive warmth is both wasteful and harmful.

As soon as seedlings are large enough to handle safely, lift them gently with a small wooden plant label or old table fork, holding them by the seed leaves, *not* the stem, and plant or prick them out with a wooden dibber in deeper trays of compost, at 24 to 60 per tray. Generous spacing gives stronger plants. Alternatively, pot them separately in 7.5cm (3in) pots.

Hardening off All young plants to be planted outdoors need hardening off to accustom them to the harsher conditions without a sudden shock. Move them first to the coolest part of the greenhouse, then to a frame outdoors, giving gradually more ventilation until after a couple of weeks they are completely exposed.

Ferns from spores Ferns, such as pteris, are raised similarly from their minute spores, which look like fine brown dust and occur on the underside of the fronds. Use pots of peat, or loam mixed with coarse grit, with a fine even surface. Water the pots, drain, then sow the ripe spores thinly, but do not cover with soil. Put a sheet of glass over the pots and stand in shallow trays of water in a warm closed propagator, shaded from sun. When the mass of tiny plants forms, separate it into small patches and transfer to well drained half pots; keep them warm and moist. When baby ferns develop, it is time to pot them singly in small pots – 7.5cm (3in) in diameter.

ABOVE Half-hardy bedding plants, like French marigolds, are raised in spring indoors.

LEFT Marigolds, and many other seeds, are sown thinly and evenly in seed trays. Cover lightly with seed compost.

PLANTS FROM CUTTINGS

There are many different types of cutting used for propagation purposes. They may be taken from the shoots, leaves, leaf buds or roots of mature plants; all are incomplete, however, as they lack some of the parts needed for an independent existence. It is the propagator's job to coax them to form the roots or shoots they need to grow and fend for themselves. When fully developed, plants raised from cuttings will be exact replicas of the parent.

Stem cuttings A great variety of shrubs, perennials, trees, alpines and pot plants can be increased from stem cuttings, of which there are several kinds. Known as softwood, semi-ripe or hardwood cuttings, they are distinguished by the woodiness of their stems, which depends on their stage of development and the time of year at which they are taken and rooted.

Softwood cuttings These are mostly taken in spring and early summer while growth is still soft. They need some heat (around 13°C/ 55°F) to form roots. Use the soft sideshoots of shrubs and alpines, but for perennials take 5cm- (2in-) long basal shoots with a little older growth at the base.

Collect cuttings only from healthy plants, while they are firm and full of water in the cool of the day, and put them in a polythene bag so they don't wilt.

Prepare them with a sharp knife, making clean cuts that are quick to heal – shrub cuttings should be about 7.5cm (3in) long, those of alpines 2.5–3.5cm (1–1½in). Strip the lower leaves, which would otherwise be buried and rot, and cut across the stem just beneath a leaf joint. Dip the end in a hormone rooting powder such as 'Keriroot', shake off any excess, then insert with a small dibber in an equal parts mixture of moist sphagnum peat and coarse sand or grit in small plastic pots or 5 cm- (2 in-) deep seedtrays.

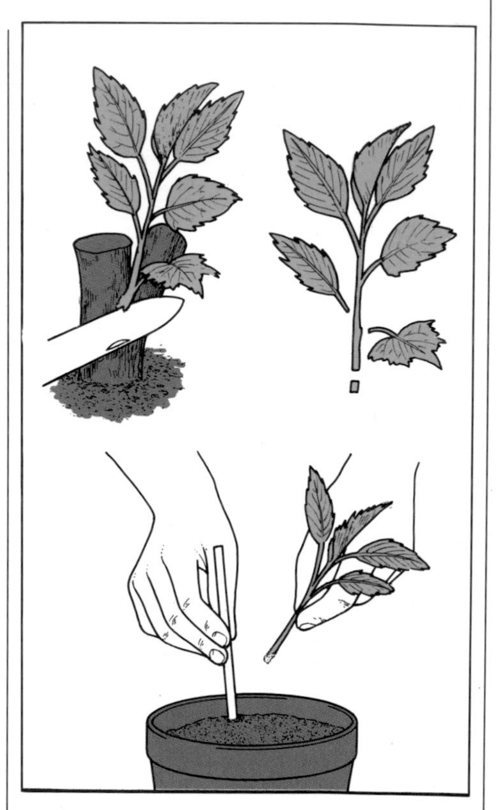

Firm in gently and water with a rosed can, then move to a warm (13–15°C/55–60°F), humid place to root. An electrically heated propagator or bench equipped with soil-warming cables will provide the necessary bottom heat – or, if you are rooting a lot of cuttings, a mist propagation unit will be helpful.

Such cuttings can also be rooted in pots on a warm windowsill. Cover the pots with polythene bags to retain humidity, but ventilate occasionally to clear condensation. Make sure that the bag is not in contact with the leaves of the cuttings.

To succeed, soft cuttings must have enough warmth and moisture, but not too much, or they will rot. Use clean containers and sterilized compost and always take care to remove any dead leaves promptly.

Mist propagation
This is particularly useful for slow-rooting cuttings such as those taken from evergreens. The unit consists of a propagating case or frame fitted with soil-warming cables to provide bottom heat and a special spray unit which sends out frequent mists of water. This cuts down on water loss from the leaves and speeds up rooting.

When shoot tips start growing strongly, remove the cuttings from the propagating frame or polythene bag, and when roots appear through the bottom of the container, pot separately in 9cm (3½in) pots of a peat-based compost such as 'Keri-compost' or John Innes No. 1 compost. Keep the plants warm until they become established, then harden them off by placing them in a cold frame and acclimatizing them gradually to cooler conditions.

Rooting in water Soft stem cuttings of indoor plants, such as impatiens (busy lizzie) and tradescantia, can be rooted in water on a warm windowsill in spring or summer. Cut 7.5–10cm (3–4in) long shoots beneath a joint, strip their lower leaves and stand in a small jar, just covering the stripped stems with water, plus some charcoal to keep it sweet. Pot them into individual 7.5cm (3in) pots when their roots are 2.5cm (1in) long.

LEFT Soft basal cuttings of dahlias should be treated with hormone rooting powder, then inserted in a peat and sand compost.

RIGHT Spring-rooted dahlia cuttings soon make large plants which will flower during the summer of the same year.

TOP RIGHT Taking a semi-ripe conifer cutting with a heel.

BELOW RIGHT This heel should be trimmed cleanly with a sharp knife.

TOP FAR RIGHT A well-rooted African violet leaf cutting – plantlets can be seen.

FAR RIGHT Plantlets growing from peperomia leaf cuttings.

Semi-ripe cuttings Taken from mid-summer to early autumn, these have firmer stems just becoming woody. They should be between 5–10cm (2–4in) long, except in the case of heathers when 2.5–5cm (1–2in) is a better length. Conifer cuttings must have at least 13mm (½in) of ripe brown wood at the base. All semi-ripe cuttings are prepared in the same way as soft cuttings but it is not essential to provide heat for rooting.

Insert semi-ripe cuttings in peat/sand cutting mix in trays or pots with a dibber, firm and water in; or plant 7.5cm (3in) apart each way in a 7.5–10cm (3–4in) layer of cutting compost in a cold frame or under a polythene cloche. All will root more quickly in heat of around 18°C (64°F) and some, such as elaeagnus, demand this degree of warmth.

Some semi-ripe cuttings root in a few weeks, others take several months, according to type, when they are taken and how much warmth is provided for them. Pot those rooted in heat and harden them off. Leave those rooted in frames undisturbed until the next spring, but making sure that they do not dry out.

Many shrub cuttings are taken with a 'heel' of older wood by pulling off a suitable sideshoot so that a strip of wood from the parent stem comes with it. Trim this strip back cleanly to leave about 13mm (½in) of older wood and insert the cuttings in compost in the usual way.

Hardwood cuttings These are taken in late autumn/early winter from shoots of the current year's growth which by now should be woody and well ripened. Select pieces 15–23cm (6–9in) long, cut just beneath a bud at the base and just above a top bud if the tip is soft.

Insert cuttings from half to two-thirds their length in a V-shaped trench in the open ground in a well-drained, sheltered part of the garden. Trickle a layer of sharp sand along the bottom of the trench to encourage rooting. Replace soil, firm well and firm back any that get lifted by frost or they will not root. Hardwood cuttings can also be rooted in half peat/half sand mixture in a cold frame or in 13cm (5in) pots in a warm greenhouse. The latter method is absolutely essential for wisteria and fruiting fig.

Don't lift cuttings rooted outdoors until the following autumn, but keep them watered and weeded.

Redcurrant and Gooseberry Cuttings

When making hardwood cuttings of these fruits, remove all but the top three or four buds so that they form 'bushes' on short clean stems.

ABOVE Hardwood cuttings out of doors are inserted in a sand-lined V-shaped trench in a sheltered well-drained spot, then firmed in thoroughly.

CENTRE If desired, streptocarpus leaves can be cut into sections, each of which will form a new plant.

ABOVE The sections are then inserted vertically into a peat and sand cutting compost.

Leaf cuttings A handy means of increasing some greenhouse plants, but they need a temperature 18–21°C (64–70°F) and high humidity to root. There are two kinds. For saintpaulias (African violet), peperomias and the rock plant ramonda, use leaves with their stalks. Cut them from the parent with a sharp knife, dip the stalk in a hormone rooting powder such as 'Keriroot' and insert upright with a dibber in a peat/sand mixture. *Begonia rex* leaves, without stalks, are laid face up on the surface of peat/sand compost after the main veins on the back of each leaf have been cut in several places. Anchor the leaf with some small pebbles. New plantlets will develop from the cuts.

Streptocarpus and gloxinia leaves are inserted upright with the top half of each leaf removed to reduce moisture loss. Pot all rooted plants in 7.5cm (3in) pots of John Innes No. 1 or one of the peat-based composts such as 'Kericompost'.

Leaf bud cuttings These consist of a leaf and short section of stem with the bud in the axil or angle between them. Take ivy and clematis leaf buds from the soft growth in spring. Cut the shoots just above a leaf bud at the top, and make the bottom cut about 19mm (¾in) beneath the leaf. Always reduce pairs of clematis leaves to singles.

Dip the base in a hormone rooting powder such as 'Keriroot', insert the stem in cutting compost with only the leaf showing and water in. Root in a warm humid place, but ventilate occasionally. When the bud shows signs of growth, the cutting will be rooted and can be potted up and hardened off. This may take from a few weeks to several months.

Use semi-ripe wood in August for camellia leaf bud cuttings. These are not easy to root but respond to a temperature of 18°C (64°F) applied as bottom heat.

Leaf bud cuttings from young growth of *Ficus elastica* and dracaena can be rooted in spring. Support ficus with a cane after rolling the large leaf and carefully securing it with a rubber band.

Eye cuttings A form of leaf bud cutting used to propagate fruiting and ornamental vines. They need bottom heat at a temperature of 21°C (70°F). Use well hardened one-year-old wood cut into 2.5–3.5cm (1–1½in) sections in mid-winter. Make the top cut just above a bud, the lower between buds. Leave only one bud at the top, excising the other, and insert upright in compost.

Alternatively, cut 3.5cm (1½in) lengths with a bud in the centre, remove a sliver of wood from the opposite side, then press horizontally into the compost leaving only the bud visible above the surface. Leave until well rooted in early spring before potting. (Don't be misled by early shoot growth.) Support young growth with a cane and, when large enough, harden off and plant out.

LEFT Camellias can be propagated from leaf bud cuttings, consisting of a leaf and piece of stem.

ABOVE Camellia leaf-bud cuttings being inserted. They need a basal temperature of 18°C (64°F).

LEFT Many perennials and shrubs can be increased from thick root cuttings. Make a slanting cut at the base.

ABOVE Pinks can be propagated from pipings. Just pull out the tops of young shoots with at least 3 pairs of leaves.

Pipings These are shoots used to propagate pinks in mid- to late summer. Just pull out the tops of young shoots with three or four pairs of mature leaves after the lowest have been stripped. Root in 9cm (3½in) pots of sand in a close, humid cold frame. Ventilate freely after 3 weeks then later, when well-rooted, pot separately in 9cm (3½in) pots of John Innes No. 1 or a peat-based compost such as 'Kericompost'.

Root cuttings A simple and reliable way of reproducing some plants is by taking root cuttings in midwinter. Obtain suitable roots by either lifting small plants or by removing soil from round larger ones to expose the roots, and replacing the soil and firming the plants back afterwards. Vigorous, young roots grow best. Cut roots into sections and root in peat/sand cutting compost, usually without heat.

Thicker roots, such as those of anchusa or eryngium, should be as thick as a pencil, thinner ones – *Phlox paniculata*, for example – about 3mm (⅛in) in diameter. Keep the cut roots moist in a polythene bag, then prepare cuttings indoors. Cut the thick type into 5cm (2in) lengths with a sharp knife, straight across the top but slanting at the base to help you to identify which is the right way up.

Use no rooting hormone but dust with captan or Benlate + 'Activex' to protect the cuttings from rotting. Then insert them upright to their full length 5cm (2in) apart in cutting compost. Lay thinner roots, such as those of phlox, horizontally 2.5cm (1in) apart on the compost and in both cases cover with a further layer of compost. Pot off cuttings in seed trays as soon as they are well rooted, but leave those that have been cultivated in frames until the autumn.

DIVISION

Making new plants by dividing up established ones is one of the most straightforward of all propagating techniques. At its simplest, division is just a matter of splitting up clump- or mat-forming plants into strong-growing pieces well equipped to start a life of their own. The little divisions will be complete plants and be provided with shoots, buds and roots – a major difference from all other methods of propagation. The methods used vary in detail according to the type of growth of the plant in question, as detailed on these pages.

LEFT Pyrethrums should be divided immediately after flowering.

BELOW Large clumps of perennials can be divided with two forks.

Herbaceous perennials A great many, though not all, herbaceous plants form spreading clumps within a few years, which eventually deteriorate in the centre and flower less well. It pays to lift and divide them every three or four years, replant young, vigorous pieces from the outer part of the clump and discard the rest. This work is often done in autumn, but the young plants – asters and heleniums, for example – stand a much better chance of getting established in early or mid-spring when it is warmer and the spring surge of growth is on. Early flowerers, such as doronicums, pyrethrums and primulas, will be in bloom at or about this time, so do not divide them until their flowers are over.

Some herbaceous plants do not oblige by forming clumps which you can split up. You will have to look to other methods to propagate them – root cuttings for anchusa and eryngium, for instance, soft basal cuttings or seeds for lupins and the hugely popular delphiniums.

Lift clumps which you intend to divide carefully with a digging fork, retaining a good ball of soil and fibrous roots. Then shake off as much soil as possible so you can see what you are about. Large clumps can be broken up by thrusting two digging forks back to back into the centre and levering them apart. A good strong sharp knife may help for cutting stubborn roots.

After reducing the original clump to several large pieces you can, if you wish, pull these apart into separate 'crowns', each with one or two strong shoots and a healthy cluster of fibrous feeding roots. The ideal size is about 7.5cm (3in) across.

Do not let any of the pieces of plant dry out. Plant them straight away, even if only temporarily, or protect with polythene bags or pieces of sacking.

Generous-sized divisions used to refurbish a border can be planted straight back in the open ground, but the smallest divisions, designed to create the largest possible stock of the plant, will do better in deep trays of a mixture of equal parts of peat and sand placed outdoors in a sheltered spot until the plants are large enough to plant out. Keep all new plants watered in dry weather and clear of weeds.

Flag irises These, and other plants which grow from rhizomes (swollen horizontal stems), are divided in summer after flowering, but the technique is a little different. Lift established clumps and pull apart with your hands. Then select healthy young rhizomes about 10cm (4in) long, each with a fan of leaves and some strong, fibrous roots for replanting. Discard old worn-out material. The leaves can be shortened back by a half to reduce moisture loss and wind resistance. Plant the rhizomes so that they are just visible on the soil's surface.

LEFT Clumps of perennials can be divided into hand-sized portions for replanting.

ABOVE Iris divisions consist of a young rhizome with fibrous roots and a fan of leaves.

Waterlilies grow from thick tubers which can be divided in spring. Wash them clean before dividing with a sharp knife. Each division must have several strong buds and some roots.

Rock plants Other plants besides the familiar occupants of perennial flower borders can be conveniently increased by division. Many rock plants form mats of growth that root as they spread and can be lifted and divided to form new stock. Alternatively some, such as sempervivums and saxifrages, form offsets – young rosettes of growth with roots of their own – which can be removed and grown separately.

All these young rock plants are best grown on for a season in 9cm (3½in) pots of John Innes No. 1 with plenty of extra grit added to provide the open sort of compost they prefer.

Grasses Ornamental grasses, such as phalaris and miscanthus and the familiar pampas grass (*Cortaderia*), form clumps that can be divided in exactly the same way as herbaceous plants, though pampas does not establish too well.

Water plants Many of the pondside plants can be treated in a similar way between mid-spring and early summer, whether they are true aquatics such as *Iris pseudacorus* and typha (reed) or bog plants, such as primulas and lysichitum. In the pool itself, oxygenating plants, such as myriophyllum and elodea, can be lifted from mid-spring to early summer, cleaned of blanket weed and other unwanted material, then divided and returned to the water, where they will quickly re-establish. Waterlilies (nymphaea) grow from thick tubers which can again be divided in spring. Lift them out and wash clean before dividing with a knife. Each piece must have several strong buds and some roots.

Greenhouse plants Division is a convenient way of propagating quite a lot of greenhouse plants, too. The popular spider plant (chlorophytum) is one, orange-flowered clivia another. This is the only way of reproducing the yellow-margined 'Laurentii' form of *Sansevieria trifasciata* (mother-in-law's tongue) true to colour.

Bromeliads, such as aechmea, neoregelia and others related to the pineapple, form offsets round the original 'vase', which dies after it has flowered. These can be eased apart and potted separately.

New divisions of greenhouse plants should all be kept warm and moist until they have had time to become properly established.

Shrubs These are generally increased from seeds or cuttings, or grafted, but a number form suckers. These are shoots that arise direct from the roots: the snowberry (symphoricarpos) is a striking example; *Rhus typhina* and yuccas are others. It is a simple enough matter to lift some of these rooted suckers, carefully severing them from the parent with the aid of a sharp knife and then planting separately wherever appropriate.

Fruits Among the most popular fruiting plants, raspberries also make suckering growth. Healthy, well-rooted suckers can be lifted in autumn and planted out to form new rows of fruiting canes.

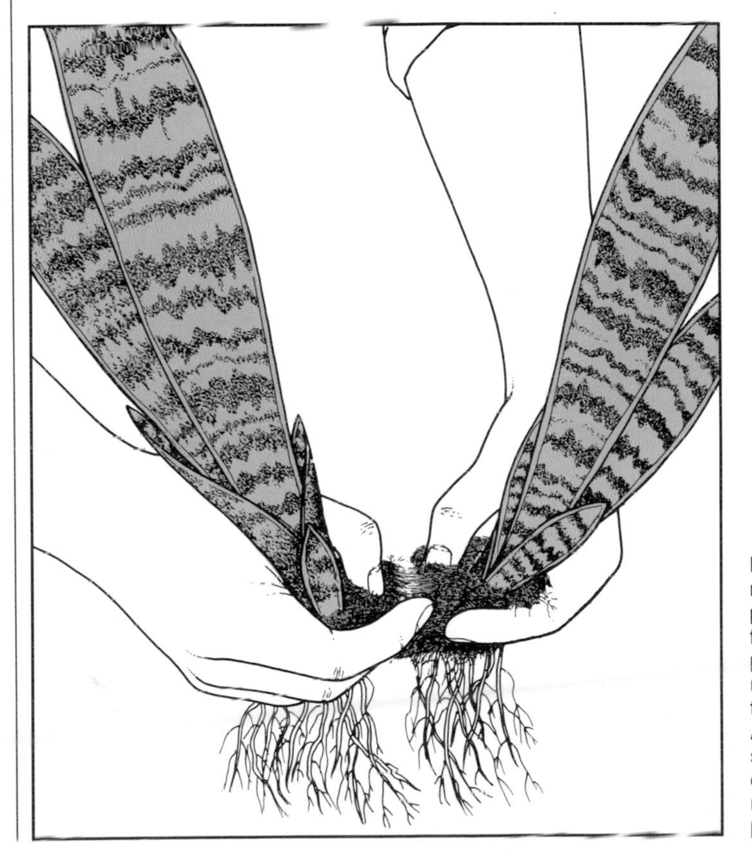

Division is a good method of propagating quite a few greenhouse plants, such as the mother-in-law's tongue or sansevieria. After potting them separately, keep the divisions warm and moist until they are properly established.

BULBS, CORMS, TUBERS

Bulb *species* can be raised from seeds, but named varieties of narcissus, tulip, gladiolus, lily, dahlia and all the other familiar kinds of bulbous plant must be propagated vegetatively if they are to be 'true to type', or identical to the parent. Apart from soft basal cuttings used to raise dahlias, the main methods are by division and offsets which, in the case of bulbs, are known as bulblets, bulbils, cormlets or daughter bulbs.

From seeds Although this is a useful way of increasing the bulb species, it usually takes 3 to 5 years to produce bulbs of flowering size. Sow the seeds outdoors as soon as ripe in drills 15cm (6in) apart and 6mm (¼in) deep. Small quantities of seeds should be sown thinly in pans or boxes and placed in a frame or greenhouse. The germination time varies, sometimes taking up to a year and, ideally, seeds should be left undisturbed for at least a year after they have germinated, after which the seedlings should be lifted and planted where required.

Bulbils These are secondary bulbs produced around the 'mother' bulbs during the growing season. Lift the bulbs in summer after flowering, carefully remove the bulbils or offsets and grow them on in a nursery bed for a year or two until they reach flowering size. This method is used for daffodils, tulips, hyacinths and many of the smaller

LEFT Lily bulbils are collected in late summer.

TOP Lily scales can be used for propagation.

ABOVE When placed in warmth, lily scales will each form a tiny new bulb at the base, when they can be transplanted.

bulbs, such as muscari. Snowdrops (galanthus) can be increased in a similar way, but this is best done just after flowering, when they still carry green foliage.

Some lilies, including *Lilium tigrinum*, *L. speciosum* and 'Enchantment', form bulbils in the axils of the leaves, where these join the stem. These can be collected in late summer, stored dry over winter, then sown like seeds (5cm/2in apart, 2.5cm/1in deep) in trays in spring to develop into mature bulbs.

Cormlets Gladioli and crocuses grow from corms, which wither after flowering having formed a new corm for the following year. But they also produce a generous crop of cormlets around the parent corm, which can be removed in autumn when the plants are lifted, stored in a cool, dry place until spring and then sown in seed trays of potting compost. It will be 2 years before they reach flowering size.

Bulb scales Lilies are a large and somewhat varied family, so they are not all propagated in the same way. A lily bulb consists of a large number of fleshy scales attached at the base. A few of these scales can be carefully removed from each bulb. (If you are prepared to sacrifice the parent bulb you could take many more scales to build up more stock.) Lift the bulb after flowering once the flower stem has died down.

Remove the scales and root them in trays of peat/sand compost or in the same mixture in polythene bags. They root better in warmth, 15–18°C (60–64°F), though this is not essential. Each scale will form a tiny new bulb at the base and they can be planted out in trays and grown on; they should then reach flowering size in two to three years.

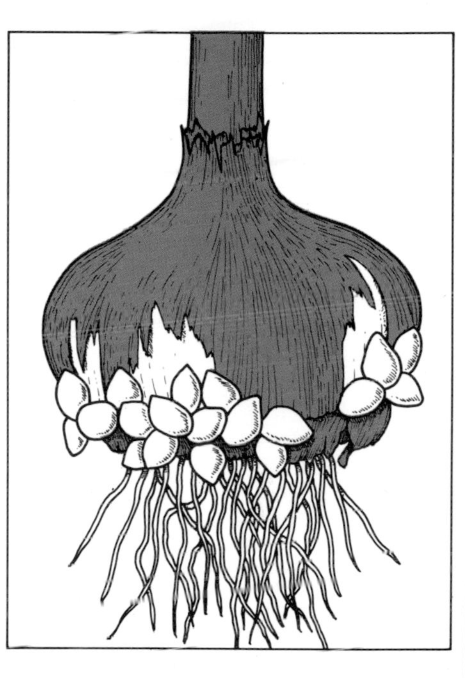

Gladiolus corms produce a generous crop of cormlets which can be removed in autumn, stored until spring and then 'sown' in seed trays of compost.

Tubers Dahlias grow from fat tubers, which are usually lifted and stored dry for the winter, then either started into growth in boxes of peaty compost in March, or planted with their crowns 7.5–10cm (3–4in) beneath the surface straight in the ground in mid-spring, so that by the time they break surface their shoots run no risk of being frosted.

Dahlias are often increased from soft basal cuttings (see Cuttings, p. 14), but the clusters of dahlia tubers can be divided while still dormant, each making several plants. Use a sharp knife and make sure each portion has one – and preferably several – strong crown buds (where stems and tubers join). Dust the cuts with flowers of sulphur or Benlate + 'Activex' to protect from rotting, and then grow them on or plant them out in the usual way.

LAYERING PLANTS

This is a method of propagation in which a stem is encouraged to produce roots so that a new plant is formed while it is still attached to, and nourished by, the parent plant. Some shrubs, such as *Forsythia suspensa* and *Jasminum nudiflorum*, often layer themselves naturally – a gift to the gardener, who can lift them with plenty of healthy roots already formed, cut them from the main plant and plant them in a chosen place.

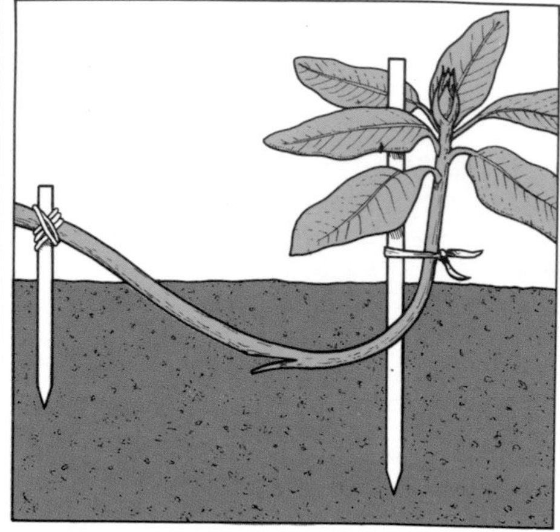

LEFT *Cornus alba* and varieties are good subjects to increase by layering.

ABOVE The stem to be layered must first be wounded to encourage rooting.

Another group that layers itself naturally (and can also be layered deliberately) is the bramble or rubus, both the cultivated blackberry and the ornamental kinds. The tips of the shoots of these plants arch over and root into the soil.

To ensure success with the simplest form of layering: select vigorous, flexible young branches, set them going at a favourable time of year (April–August), and prepare the soil well round the layer by digging it over and mixing in plenty of peat and sand to encourage the wounded tissue to produce a healthy root system.

The stem chosen should be notched or cut obliquely halfway through at a leaf joint about 30cm (1ft) from its tip. Treat the cut surface with a rooting powder such as 'Keriroot' and wedge the cut open with a sliver of wood. Peg down the shoot so the cut area is touching the soil and stake the tip of the shoot upright so it forms a more shapely plant. Water in dry weather and keep clear of weeds. Border carnations can be layered in a similar way, too, in late summer after flowering.

Serpentine layering This method follows the same principles, but here a long shoot of a jasmine, honeysuckle, clematis or vine is layered in several places so that it will yield a number of new plants.

Runners Strawberries (and some other plants, such as chlorophytum and *Saxifraga sarmentosa*) increase themselves naturally by means of young plants that form on the runners which spread from the plants in summer. To conserve the plant's energies for fruiting it is usual to remove all runners not required for propagating.

The first plantlet to form on each selected runner is pegged into a pot of potting compost plunged in the ground by the plant. As soon as it is rooted it can be severed from the parent and grown on separately. The remainder of the runner should be cut short. Layer strawberries in early and mid-summer, keep them moist and lift the potted plants in late summer and early autumn. Be careful not to propagate diseased plants, especially those that have been infected by virus.

Air layering Once called Chinese layering, because of its origin, this can be used where stems cannot be bent to the ground. The stem, whether of a rhododendron outdoors or a rubber plant (ficus) indoors, is cut in the same way as in an ordinary layer, treated with rooting hormone, the wound propped open, and the whole area wrapped in moist sphagnum moss and bandaged in clear polythene sheeting. Within a couple of months roots will form and the rooted part of the plant can be severed and potted up separately.

Air layering a rubber plant. The wound is held open and then wrapped in moist sphagnum moss.

The moss is held in place with a 'bandage' of clear polythene sheeting which keeps it moist.

GRAFTING AND BUDDING

The purpose of grafting is to combine the vigour (and perhaps disease resistance) of a rootstock with the choice fruiting, flowering or decorative foliage of another plant, the scion. The rootstock and scion must be closely related for the graft to take successfully and are usually taken from plants belonging to the same genus.

Guidelines for success Before describing grafting and budding techniques, it is important to grasp the principles underlying this kind of propagation. As the object of the exercise is to marry parts of two plants to produce a new one with the best characteristics of both, the first essential is to choose the right pieces. For example, to graft the apple 'Cox's Orange Pippin' it is necessary to have a suitable scion from a tree of that variety. The rootstock must be compatible with it – that is, another apple – or the two will not knit together. A rootstock of the right vigour for the purpose must also be chosen – Malling 9 for a small compact tree for a tiny garden, MM106 for something larger. The same goes for ornamentals. Choice laburnum hybrids are grafted on to common laburnum (*L. anagyroides*), rowan cultivars on to *Sorbus aucuparia* and roses on to a rose stock, usually *Rosa canina* for bush roses and *R. rugosa* for standards.

Rootstocks for ornamentals are generally raised from seeds. If you decide to grow your own, set out the seedlings in a nursery bed to grow on until they are two years old and ready for grafting. They will need to grow on for a further two years before they are large enough to set out in the garden. Fruit tree stocks must be bought in.

Two things are essential if the two

Apples, like 'Cox's Orange Pippin', must be grafted on to apple rootstocks of the right vigour for the type of tree that it is desired to produce. Very often these days dwarfing rootstocks are used to produce small or trained trees.

Whip and tongue grafting. At far left, two angles of the stock and scion cuts are shown. Near left, the scion has been inserted and tied in.

parts are to grow as one: to make accurate cuts with a grafting or budding knife, and to fit together the cambium layers in stock and scion. The cambium is the actively growing layer of green tissue just beneath the bark, whose cells are multiplying all the time.

If grafting or budding is done correctly, the stock and scion tissue will fuse to form a plant.

Other essential precautions are to protect the graft or bud from drying out until it has taken, and from the entry of disease. Wrap it round with raffia or grafting tape. Release this later when it is no longer necessary.

Whip and tongue This type of graft is commonly used for propagating fruit and ornamental trees where the stock and scion are of approximately the same size. The job has to be done outdoors in March, so it is wise to choose a reasonably warm day for it; it isn't possible to make accurate and safe cuts with frozen fingers!

To carry out a whip and tongue graft, first cut back the rootstock to 10cm (4in) above soil level, then make a sloping cut 5cm (2in) long at the top. Now form the tongue by making a short downward cut in this sloping surface, using a sharp knife.

Take suitable scion wood from the tree you wish to propagate; choose vigorous, well-ripened one-year-old shoots, each with four buds. Cut just above a bud at the top and 2.5cm (1in) below a bud at the bottom. Now make a slanting cut at the lower end of the scion to match the prepared stock and with a tongue as shown in the diagram.

Fit the two together snugly by pushing the scion's tongue into that on the rootstock, then tie the graft firmly in place with raffia and paint round with grafting wax to protect it from disease and the weather. When the scion starts to grow, carefully cut through the raffia.

Saddle graft This is well named (see illustration) as the scion sits astride the stock just like a rider on horseback. This graft is used mainly for propagating large hybrid rhododendrons, which are difficult to strike from cuttings. *R. ponticum* is used as the stock. It can be raised from seeds but needs to be two or three years old before it is substantial enough to graft. Semi-ripe cuttings of *R. ponticum* can be rooted with 21°C (70°F) bottom heat, but are difficult.

Move the potted stocks into a warm greenhouse several weeks before grafting to get them growing, as these rhododendron grafts need heat to take. Make sure the rootstock and scion are of the same diameter to create a good union. Cut back the rootstock to 5cm (2in), then form a wedge to take the scion by making two small, upward sloping cuts about 2.5cm (1in) long.

Cut scions 10cm (4in) long and make two upward cuts in the base of each to form a notch that will fit the saddle. Fit the scion, bringing the cambium layers of the two parts exactly together, then bind round with raffia and paint with grafting wax to keep out wet and disease.

Stand grafted plants in a propagator so that they receive the necessary bottom heat in order to take. They should knit together in about six weeks, after which the raffia must be cut so that it doesn't restrict growth. Then gradually harden off the plants ready to set them out.

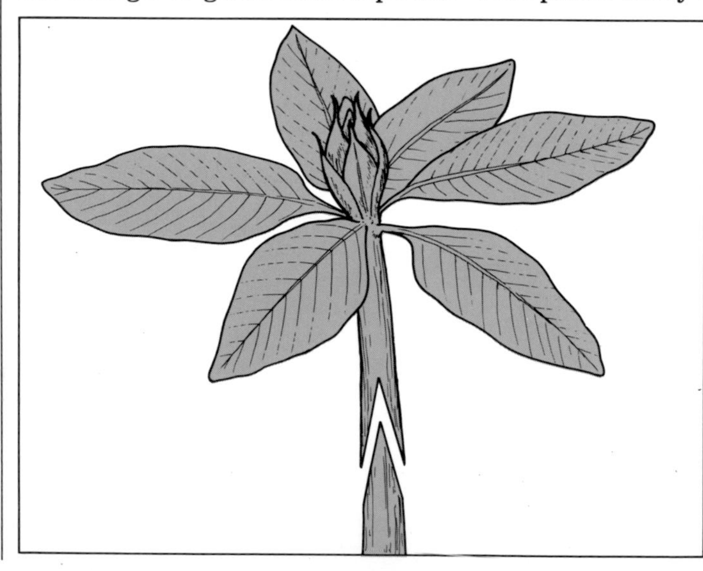

The saddle graft is used mainly for propagating the large hybrid rhododendrons, which are often difficult from cuttings. The scion sits astride the stock. Two- to three-year-old plants of *R. ponticum* are used as stocks.

Dwarf and larger conifers can be propagated by grafting, using the veneer graft. First a flap of bark is cut in the stock. Then the base of the scion is cut to a thin wedge shape and then pushed behind the flap of bark. Finally, the graft is tied firmly in place with raffia.

Veneer graft The wide range of dwarf and full-sized conifers with attractive foliage of many colours and textures is also generally grafted. Here the veneer graft is the one to use.

Chamaecyparis lawsoniana and *C. obtusa* cultivars, for example, are grafted on to ordinary *C. lawsoniana, Picea pungens glauca,* the blue spruce, on to *P. pungens* or *P. abies*. Golden yew goes on to the common yew.

Raise rootstocks from seed sown outdoors and select for grafting when they are two years old. Pot them in 9cm (3½in) pots and take into a heated greenhouse several weeks before grafting. Leave the top on the stock and make a shallow vertical cut about 2.5cm (1in) long near soil level to form a flap of bark with wood and cambium exposed.

Select scion wood of the previous year's growth about 10cm (4in) long (though perhaps half that for dwarf conifers). Cut the base of the stem to form a thin wedge of the same length as the cut in the stock.

Push the scion into the flap of bark, matching the cambium layers, then tie with raffia. Move the grafted plants into a warm propagator for about six weeks to knit together. After the graft has taken, cut away the top of the rootstock and gradually harden off the plants ready for planting outdoors.

Budding Budding is to grafting what bud cuttings are to stem cuttings – the objective is the same but a smaller part of the plant is used. As the starting-point is only a bud, it will take longer to produce a mature plant by this method.

Budding is used primarily for producing named varieties of rose – hybrid teas, floribundas, miniatures, climbers and ramblers. As with grafting, the result is a combination of the vigour of the rootstock and the choice flowers of the scion.

Fruit and ornamental trees can be budded as well as grafted and this certainly has the advantage that it can be done during the warmer summer weather.

If you want to bud your own roses, your first step is to obtain suitable rootstocks. *Rosa canina*, the dog rose, is widely used and can be raised from seeds sown outdoors. These will need cold treatment to break their dormancy before they will germinate (see page 9). Allow about 18 months from the time of

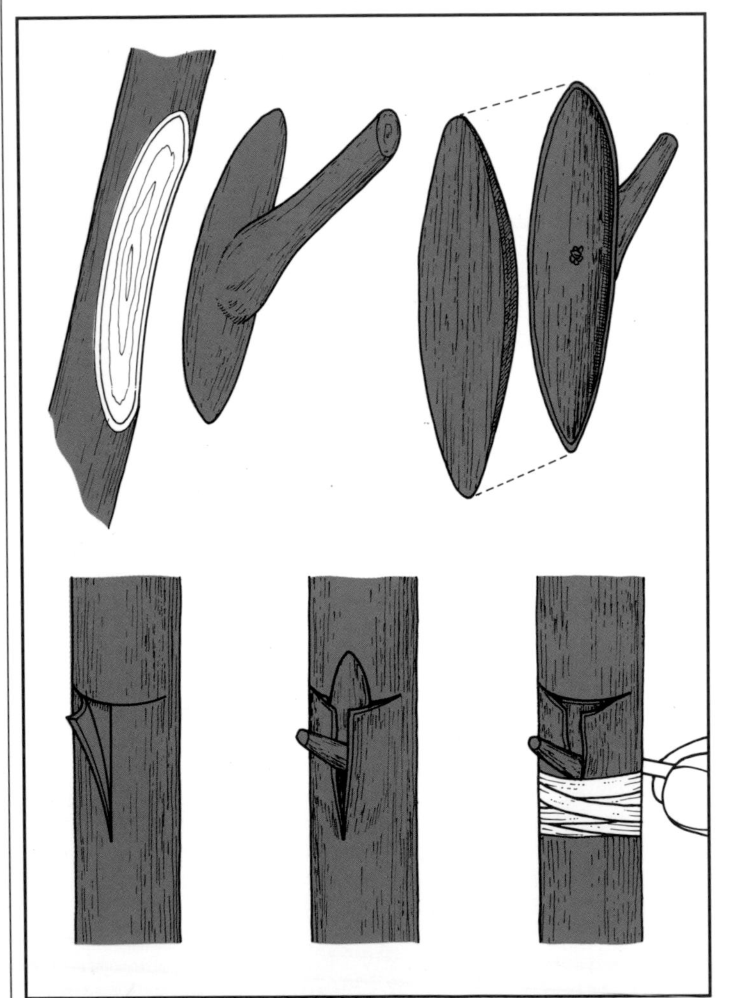

Budding is used primarily for producing named varieties of roses. The bud is removed on a shield-shaped piece of bark and a T-shaped cut is made in the bark of the rootstock. Then gently remove sliver of wood behind the bud. Insert bud in the T-cut and tie in with raffia.

Most roses, such as the hybrid teas (shown here) floribundas, miniatures, climbers and ramblers, are best propagated by budding as the resultant plants are more vigorous.

sowing before you get usable plants.

Plant out seedling rootstocks 30cm (1ft) apart in a nursery bed in autumn. The stem should stand above the soil, but earth it up in winter to keep it soft, ready for budding, when the soil is removed.

To bud a standard rose you need a different stock – *Rosa rugosa* 'Hollandica'. This can be raised from hardwood cuttings (page 16). Remove any buds likely to be buried, or you will get many unwanted suckers. Train a single stem 90–120cm (3–4ft) up a cane to keep it absolutely straight and insert buds of the desired scion when the stem is two years old.

The bark is easily lifted in summer, provided the plant is not short of water, so water beforehand if necessary. Then select from the variety required suitable current year's shoots with plump buds. Remove the soft tip and all the leaves, but *not* the leaf stalks, because these will be useful later.

Next make a shallow T-shaped cut in the rootstock close to soil level, 2.5–3.5cm (1–1½in) vertically, and 6–13mm (¼–½in) horizontally. Gently lift the bark on each side of the vertical cut with the spatula on your budding knife and insert the prepared bud immediately, so that it doesn't dry out.

To prepare buds, insert your knife 13–19mm (½–¾in) below a bud, cut upwards behind it to come out the same distance above it. This gives you a bud in a shield-shaped piece of wood held by the leaf stalk. Gently remove any sliver of wood from behind the bud, then insert the bud in the T-cut. Trim off any surplus bark protruding above the T-cut, then tie with raffia. Cut off the portion of rootstock above the bud the following spring, and remove the ties when growth starts.

Bud standard roses high on the stem where you want the head to form. For these it is usual to insert three buds spaced round the stem.

TOOLS AND EQUIPMENT

You don't need a lot of costly tools to start propagating. In fact, the less ambitious techniques require few. Always buy the best quality, however – especially knives. A budding or grafting knife should be designed for the purpose and be made of good steel, so that it stays sharp and has a long life. Invest in an oilstone, too, to keep it sharp. Many propagators prefer to use a scalpel for soft cuttings.

You will need a stock of pots and trays, plastic being easier to keep clean. The 7.5cm (3in) and 9cm (3½in) pots and 12.5cm (5in) half-pots are the most useful, with 5cm (2in) deep trays for seed-sowing, and 7.5cm (3in) for cuttings and pricked-out seedlings.

A dibber is vital but is usually homemade from a piece of 13mm (½in) thick dowelling, sharpened to a blunt point. Another home-made tool is a wooden presser to firm and level compost in seed trays.

A wooden label will do to lift seedlings for pricking out, but an old table fork is better. For air layering you will need sphagnum moss and polythene sheeting, raffia for grafting, and polythene bags to collect cuttings and keep them moist until they are inserted.

While many seeds and cuttings can be started on a windowsill, a heated propagator with a temperature range up to 21°C (70°F) is necessary for the more demanding plants, or if you want to start early in the season. A small greenhouse or a heated frame is superior to a windowsill as it gives better light and stronger plants.

For seed-sowing outdoors (and for hardwood cuttings) a garden line and measuring rod will help you to keep things shipshape.

A hand sprayer is essential for misting cuttings and moistening seedlings and it is a good idea to keep a second sprayer solely for use with fungicides and insecticides.

Finally, hormone rooting powder (or liquid, if you prefer) has become indispensable for getting the highest 'take' of cuttings.

Composts The propagator's needs in the way of composts and fertilizers are fortunately modest, but always go for reliable, good quality products.

One of the best ways of ensuring a good germination of seeds is to use a reliable seed-sowing compost – either the John Innes soil-based kind or a peat-based product such as 'Kericompost'. This provides the right open, yet moisture-retentive texture which creates the most favourable conditions for germination, and the necessary nutrients – phosphates in particular – to get the seedlings off to a good start.

Once they are growing well, you'll need to prick them out to give them more space and an additional supply of nutrients. Use John Innes potting compost No. 1, or a peat-based potting compost such as 'Kericompost'.

If you have to deal with rhododendrons, camellias or heathers, you will need to choose your composts more carefully, as these and a few other plants cannot abide lime or chalk in the soil. Look for an acid or ericaceous compost, which is free from lime.

Cuttings respond best in a 50:50 peat and sand mixture which you mix yourself. Buy a good brand of

One does not need a lot of costly tools to start propagating: pots, seed trays, a dibber, knife, simple propagator and watering can will take care of most needs.

sphagnum moss peat and suitable sharp, gritty sand. Some garden centres sell suitable sand, others will refer you to your local builder's merchant – fine, provided you don't come away with soft sand, which is suitable for mortar but not gritty enough for propagating. Make sure it's clean too – and from a source not likely to contain weed seeds.

Some cuttings – pinks particularly – root most quickly in pure sand, but be quick to move them into something more nourishing once they have sufficient root growth to be able to benefit from it.

Water supply As for watering, don't imagine water collected from the greenhouse roof is ideal. It may be soft, but it could be full of disease spores and weed seeds. You would be wiser to use clean tapwater for sensitive young seedlings. Rhododendrons, *Primula obconica* and *P. malacoides* among others won't be happy if tapwater is limy, so collect rainwater for them as clean as possible and store it in a butt with a lid.

PESTS AND DISEASES

All your skill at seed-sowing and rooting cuttings could count for nothing if you don't protect your young plants from their enemies – that is, pests and diseases. You must be able to recognize those enemies and know what weapons to use to beat them.

LEFT Never propagate from plants which have virus, such as mosaic (yellow mottling).

ABOVE Botrytis or grey mould affects not only established plants and flowers, but also cuttings of all kinds.

Diseases Three types of disease particularly need watching for when propagating. First, viruses. Examine your raspberries, strawberries, dahlias and chrysanthemums above all for any early signs of yellow streaking (mosaic) in their foliage, unusual or distorted leaf shape or stunting, all signs of virus infection. If you find any, burn them: *never* propagate from them.

Another devastating disease is damping off, a group of related fungal diseases that attack young seedlings at soil level. The seedlings then rot, topple over and die. Infection can travel like wildfire. So use sterilized compost, pots and trays, and ventilate adequately to avoid the stagnant air that encourages this trouble. Spray with Cheshunt Compound when first seen.

Botrytis or grey mould is another common fungus trouble, which is usually worst in winter. Counter it in a similar way, and by spraying with Benlate + 'Activex'.

Pests Aphids (greenfly and related pests, sometimes known as plant lice) can transmit virus diseases as they feed, quite apart from weakening plants by sucking the sap from them, so keep watch for them. Act promptly, controlling them by spraying when necessary with an insecticide such as 'Rapid' or 'Sybol'.

When preparing cuttings, especially softwood ones which are prone to attack by virus, take additional precautions and make sure all equipment is clean.

Red spider mite often appears on pot and greenhouse plants, so make sure your propagating material is clear of it. Symptoms are fine pale yellow speckling on the leaves. In bad attacks tiny spider's webs may also be seen. Spray with an insecticide such as 'Sybol', if this proves to be necessary.

HOW TO MAKE MORE

There is not space here for a directory of all the popular plants and how to propagate them, so the methods used for each main category of plants are indicated, with a few examples. (The techniques referred to are described in more detail earlier in the book.)

Lewisia can be increased by division in the spring.

Alpines

Many rock plants (species and good seedsmen's strains) such as aubrieta, phlox, aquilegia and leontopodium (edelweiss) are raised from seeds sown outdoors (perhaps in pans) in January/February. Sow gentians later, in May/June. Spreading kinds such as antennaria, campanula, lewisia and thymus can be divided in the spring (March/April). Various kinds of cutting are used: softwood from April to June for aethionema, aubrieta and veronica; semi-ripe in June and July for phlox and helianthemum; hardwood in October/November (in frames for aubrieta); leaf cuttings (in a frame or greenhouse) in June/July for ramonda, haberlea and lewisia. Saxifraga and sempervivum offsets are often already rooted or easily struck.

Annuals

These are all raised from seeds, but timing and methods of sowing vary.

HARDY ANNUALS These (e.g., nigella, clarkia, linum, calendula) are sown outdoors in March/April where they are to flower, in patches or in rows for cutting. The hardiest (e.g., calendula, godetia, cornflower) can be sown *in situ* in September or early October to give an earlier spring display. Many hardy annuals can be sown at similar times, in pots, to bloom in an unheated greenhouse.

HALF-HARDY ANNUALS Plants such as ageratum, scarlet salvia, tagetes are sown in pots or trays in warmth between January and April, then pricked out in trays or potted separately to produce summer bedding plants or decorative pot plants. Sow slow-maturing antirrhinums and *Begonia semperflorens* in January/February; most half-hardy annuals (e.g., lobelia, nicotiana, salvia, tagetes, nemesia, mesembryanthemum, petunia, cosmos) under glass in February/March; and quick growers such as zinnias in March/April.

Godetia is very hardy and can be sown in spring or autumn.

Divide the water iris, *I. pseudacorus*, after flowering.

Aquatics

Marginal plants such as typha, butomus and sagittaria are divided in spring like perennial border plants; *Iris pseudacorus* and other irises after they have flowered. Sow seeds of bog primulas when they are ripe in late summer. Waterlily (nymphaea) tubers can be divided from April to June, but each portion must have a strong bud. They can also be grown from seeds sown in a frame when ripe. Floating plants such as stratiotes provide offsets in summer. Oxygenating plants such as elodea and myriophyllum can be cut apart from April to June, whether rooted or not, and these will soon grow independently in the water.

Biennials

Wallflowers (cheiranthus), forget-me-nots (myosotis), double daisy (bellis), Canterbury bell (*Campanula medium*) and others are raised from seeds sown outdoors in a seedbed or cold frame in May/June (like cabbage plants), then set out in rows in July/August to form mature plants ready to plant in October where you want them to be in flower the following spring.

Increase aechmea from offsets.

Bromeliads
Ornamental plants of the pineapple family, such as aechmea, neoregelia, vriesia and ananas itself, are grown from seeds sown under glass in March/April, or from the rooted offsets (April–July) that form round the parent plant (which will die after flowering).

Bulbous plants
Natural species of bulbs can be raised from seeds, best sown outdoors preferably in pans, as soon as ripe or early in the new year. Most garden bulbs, however, are hybrids, so must be increased vegetatively – snowdrops (galanthus) from bulblets just after flowering; crocuses from cormlets in spring; muscari from bulblets in September; and tulips from bulblets in October/November (though watch for deterioration from virus diseases). Narcissi come from bulblets in September, gladioli from cormlets in April. Lilies can be increased from bulb scales in spring or summer or from bulblets in autumn. Some (e.g. *Lilium tigrinum, L. speciosum* and 'Enchantment') from bulbils in the axils of the leaves, which can be collected, stored dry in winter, then sown in seed trays in spring. Eranthis (winter aconite) tubers can be divided in August/September.

Increase kalanchoe from plantlets.

Cacti and succulents
Cacti are generally raised from seeds sown under glass January–April (e.g. rebutia, chamaecereus, epiphyllum) or from cuttings June to August (chamaecereus, epiphyllum, opuntia). Use offsets to propagate rebutia. Christmas and Easter cacti (schlumbergera and rhipsalidopsis) root easily from two- or three-joint stem cuttings put under glass during the summer.

The succulents, echeveria and sedum, come from seeds sown January–April, or can be divided in spring or rooted from leaf cuttings in June/July. Echeveria also gives rooted offsets in spring and summer. The tiny plantlets that form on bryophyllum leaves can be potted as soon as they are ready. Kalanchoe can be struck from stem or leaf cuttings from June to August. Hardy sempervivums can be divided in spring, alternatively you can take rooted offsets in summer.

Climbers

Annual climbers, such as eccremocarpus, *Thunbergia alata* and ipomoea, are raised from seeds in the same way as half-hardy annuals. The woody species (but not named cultivars) can also be raised from seeds, sown in a greenhouse in March/April. Cuttings of various kinds produce plants more quickly and will replicate cultivars exactly. Clematis are grown from leaf bud cuttings taken in June/July under glass, or layered outdoors between April and August. For lonicera (honeysuckle) use semi-ripe cuttings in June/July, or in a frame October/November, or layer between April and August. Wisteria is layered in summer or grown from hardwood cuttings in November/December under glass. For wall shrubs such as garrya and pyracantha, see details given under Shrubs, page 46.

Raise annual climbers, such as ipomoea, under glass in spring.

Dahlias

Dwarf bedding dahlias can be grown from a good strain of seed sown in warmth from February to April, but named cultivars must be raised vegetatively. Existing clumps of tubers can be divided at planting time in March/April, making sure each has at least one strong bud. Soft basal stem cuttings are taken in March from overwintered tubers and then started into growth in heat in February.

Ferns

All species can be raised from spores (the fern's equivalent to a flowering plant's seeds) in a greenhouse as soon as they are ripe. Those, such as adiantum, asplenium, phyllitis and pteris, that form clumps, can be divided in April, hardy kinds outdoors, others in a greenhouse. The plantlets that form on the fronds of *Asplenium bulbiferum* soon root to form healthy new ferns.

Bedding dahlias – grown from seed.

Plantlets on fronds of the fern *Asplenium bulbiferum*.

Sow cinerarias in June.

Flowering pot plants

Many (e.g., cineraria, calceolaria, various primulas, schizanthus and salpiglossis) are raised from seeds. Sowing time varies according to type and when they are needed in flower. Cinerarias are sown in June to flower December–February, primulas in August for March, or March for a summer display. Tuberous begonias can be grown from seeds or by cutting the tubers in half in spring. Fuchsias root readily from soft or semi-ripe cuttings in warmth between April and August. Named varieties of geranium (zonal, regal and ivy-leaved pelargonium) are commonly raised from half-ripe cuttings under glass in August, but many superb varieties are now raised from F1 seeds sown in warmth in January. Florist's cyclamen are grown from seeds sown in warmth between August and March, according to variety. Leaf cuttings are the main way to increase saintpaulias, streptocarpus and gloxinias. Lily-like clivias form strong clumps which can be divided in spring to make more plants.

Foliage pot plants

Grevillea, jacaranda and eucalyptus are raised from seeds in warmth in January–April, and coleus, too, from good strains of seed, though its named forms are grown from soft stem cuttings in spring or summer. Some kinds, such as aglaonema, form clumps, others, e.g. fittonia, spread and can be divided in April. Peperomias can be divided in April or rooted from leaf cuttings in summer. Chlorophytum forms plantlets on its flowering stems in summer which can be rooted in small pots, then grown on. *Ficus elastica* can be grown from semi-ripe or leaf bud cuttings in April–July and leggy plants can be refurbished by air layering. Monstera (Swiss cheese plant) grows from stem or leaf bud cuttings in June–August. Poinsettias, grown for Christmas, are rooted from soft stem cuttings in May/June. Tolmiea, the pig-a-back plant, can be divided in March/April, or the plantlets on its leaves can be rooted in a greenhouse June–August, or the stems can be layered in spring or summer. The plantlets of *Saxifraga sarmentosa* can be rooted in small pots like strawberry plants.

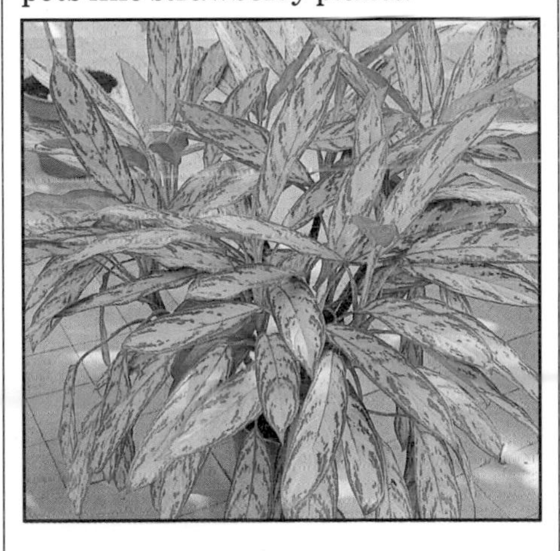

Divide aglaonema in April.

Gooseberries are easily raised.

Fruit canes and plants

Blackberries, loganberries and other brambles are reproduced by means of tip layers in July/August. Raspberries come from rooted suckers from virus-free stock plants outdoors in November/December.

Strawberries form plantlets on their runners in mid-summer which can be rooted in small pots plunged near the parent plants. Sever from the parents when well rooted and again avoid virus-infected stock.

Melons are grown as annuals from seeds sown in warmth in April, then potted off or planted in a frame or the open ground, whichever best suits the variety.

Fruit trees and bushes

Cultivated fruit trees (apple, pear, plum, cherry, quince, etc.) are all choice hybrids, which therefore have to be reproduced vegetatively. They are usually grafted on to compatible rootstocks in March, or budded in June/July.

The bush fruits, black, red and white currants and gooseberries, are struck from hardwood cuttings outdoors in November/December. Leave buds on the buried portion of blackcurrant cuttings so that they form stools, with fresh fruiting wood produced from under the soil each year.

Grow mint from rooted pieces.

Herbs

A diverse group of annuals, biennials, perennials and shrubs, increased in a variety of ways. Many (e.g., chives, parsley, dill, chervil, fennel) can be raised from seeds like hardy annuals (though parsley is a biennial) or half-hardies (e.g., basil). Chives can also be increased by dividing existing clumps.

Mint will grow from root cuttings, or rooted pieces can be taken from an existing colony. Patches of thyme and marjoram root as they spread, so rooted pieces can be lifted to make new plants. Use semi-ripe cuttings in August to make more of the shrubby sage and rosemary.

Perennials
Border plants such as helenium, achillea and delphinium are most readily increased by dividing existing clumps, or removing vigorous young rooted pieces from round the edges. This is best done in March or early April as growth starts, though plants can be divided while dormant in winter at some risk of loss. Early flowerers such as doronicum and pyrethrum are best divided just after they have bloomed.

Most perennials can be raised from seeds sown outdoors in a pan or seedbed in May or June, then set out in nursery rows to develop sizeable plants in readiness for the autumn planting season.

A useful group (oriental poppy, echinops, eryngium, anchusa, phlox) will grow from root cuttings taken in a cold frame in December/January. Herbaceous peonies are grown from well-budded divisions of their tuberous roots in March/April, though they take a year or two to recover and resume flowering. Fleshy alstroemeria roots can be lifted and divided in April, but this plant, too, resents disturbance. Divide flag irises in early July after flowering to make new plants or refurbish old plantings. Select well-rooted sections of rhizome, each with a fan of healthy leaves. Another group (e.g., delphinium, lupin, chrysanthemum and pyrethrum) can be raised from soft basal cuttings taken early in the year under glass – specially useful for choice cultivars.

Alstroemeria – resents disturbance.

Lupins can be raised from soft basal cuttings taken in spring.

Roses

Named cultivars – hybrid teas, floribundas, miniatures, shrubs and climbers – are generally budded on to suitable rootstocks between June and September. Ramblers can be rooted from hardwood cuttings outdoors in November/December. Rose species can be raised from seeds sown outdoors in March/April, though they may vary slightly from their parents. Some gardeners root hardwood cuttings of hybrid teas, but budding on to a rootstock is thought to produce stronger plants.

Shrubs

A wide range of techniques is used here. Many species are raised from seeds in spring, some of which may need cold treatment to break their dormancy. Many kinds (e.g., weigela, garrya, philadelphus, cistus, pyracantha varieties) can be grown from semi-ripe cuttings in June–August under glass. Hardwood cuttings are used for ribes, forsythia, *Buddleia davidii* and others. Grafting (saddle graft) is used for hybrid rhododendrons. Some shrubs, such as forsythia and cotoneaster, can be layered between April and August. Symphoricarpos produces suckers which can be lifted to make new plants.

Heathers (erica and calluna) can be struck from 2.5–3.5cm (1–1½in) semi-ripe stem cuttings taken from June to September. They can also be layered by working peat into the centre of mature plants so that the branches form roots and can later be detached. Seeds of erica species can be sown under glass in March/April.

Weigela – use semi-ripe cuttings.

Roses are generally budded.

Sweet peas
Raised from seeds, either sown in pots in September/October and overwintered in a cold frame until spring planting time, or sown similarly in a cool greenhouse in February/March, or sown direct in the open ground in March/April.

Sweet peas – sow spring or autumn.

Trees
Species can generally be raised from seeds, some of which (e.g. quercus, aesculus) must be sown straight away, while others (sorbus, malus, prunus) need cold treatment to break their dormancy. Choice forms such as *Acer platanoides* 'Drummondii' and *Acer palmatum* varieties have to be grafted on to suitable rootstocks in February/March to pass on their characteristics.

Ailanthus (tree of heaven) and *Rhus typhina* (sumach) grow from root cuttings taken in a frame in December/January. Rhus also forms rooted suckers which can be lifted and planted out in March/April.

Hardwood cuttings can be taken in November/December if you want to increase salix (willows) and populus (poplars).

Conifer species will grow from seeds sown outdoors in spring, but special forms – unusual shapes or leaf colours or dwarfs – are grafted in March in a greenhouse (e.g., pinus, picea, chamaecyparis) or rooted from semi-ripe cuttings in September/October in a frame (e.g., picea, juniperus, chamaecyparis).

Vegetables
Most are grown from seeds, like annuals. Hardy carrots, beetroot, lettuces, parsnips, are sown in rows outdoors where they are to crop and thinned to a suitable spacing. Tender tomatoes, cucumbers, sweet corn, peppers and celery are raised like half-hardy annuals, then hardened off and either set outdoors when frosts are past or grown in large pots, growing bags or in the ground in a greenhouse.

Cabbage, Brussels sprouts and similar plants are raised in a seedbed outdoors, then later set out in rows where they are to mature. French and runner beans and marrows are raised in pots or trays under glass, then planted outdoors, or sown direct when frost can no longer harm them. Potatoes and Jerusalem artichokes are grown from tubers, asparagus from its fleshy rootstock, raised from seed by nurserymen. The tuberous roots of rhubarb can be divided in spring; each crown should have at least one strong bud.

Vegetables – mainly seed raised.

INDEX AND ACKNOWLEDGEMENTS

Picture credits

Gillian Beckett: 1, 16(tl), 16(bl), 17(t), 18(l,r), 19.
Pat Brindley: 4/5, 7, 8, 11(t), 13(t), 20(t), 24(bl), 27(b), 28, 38 39, 40(r), 41, 42(b), 43(t,b), 44(l,r), 46(t,b), 47(b).
Nelson Hargreaves: 16(tr).
ICI: 36(l,r).
Harry Smith Horticultural Photographic Collection: 12, 13(b), 17(c,b), 24(br).
Michael Warren: 6, 10, 11(b), 15, 26, 33, 37, 40(l), 42(t), 45(t,b), 47(t).

Artwork by Simon Roulstone.